A DEMONSTRATION OF DIGITAL RADIOGRAPHY

Technique for the Bitewing Exposure (BWX) and Periapical (PA) X-Ray with Digital Sensor

Claire H. Rossi, BA, RYAN, RDAEF
Hazel O. Torres, MA, CDA, RDAEF
illustrated by David Fierstein

First Edition

Published by:

THE BITEWING™
P.O. Box 2942
Santa Cruz, CA 95063-2942

www.the bitewing.com

DEDICATION
Dedicated to the DENTSPLY RINN component collection,
and to illustrator, graphic designer, and editor David Fierstein

A DEMONSTRATION OF DIGITAL RADIOGRAPHY
Technique for the Bitewing Exposure (BWX) and Periapical (PA) X-Ray with
Digital Sensor, First Edition
Copyright 2005 THE BITEWING™
All Rights Reserved.
Notice: Neither the publisher, illustrator, editor, designer nor the authors assume any liability for injury to person or property arising from this publication.

First Edition July 2006 (USA).
Printed in the United States of America
THE BITEWING™
PO Box 2942
Santa Cruz, CA 95063-2942 www.thebitewing.com
Standard Address Number: SAN: >>> 851-3287 <<<

Library of Congress Control Number: 2006905355
A demonstration of digital radiography technique for the bitewing exposure
(bwx) and periapical (pa) x-ray with digital sensor
1st ed.
The bitewing™
Santa Cruz, CA

Claire H. Rossi, BA, RYAN, RDAEF
Hazel O. Torres, MA, CDA, RDAEF
David Fierstein, Illustrator, Editor
International Standard Book Number 0-9767786-0-2
European Article Number: EAN 9780976778608

CREDITS

Authors
Claire H. Rossi, BA, RYAN, RDAEF
Hazel O. Torres, MA, CDA, RDAEF

Editor
David Fierstein

Illustration & Design
David Fierstein www.davidiad.com

ACKNOWLEDGMENTS

Collectible Material Featured with Permission:
XCP® Evolution 2000 / XCP-DS® Digital Sensor
Courtesy of : RINN Division of DENTSPLY International, Inc
Figures 1, 2, and 6ª courtesy of DENTSPLY RINN

XDR Digital Radiography from:
CYBER MEDICAL IMAGING, INC.
Figures 12, 13, and 14 courtesy of CYBER MEDICAL IMAGING

Reviewer:

Jon M. Koeltl, DDS
Carol H. Martin, RDH
Doni L. Bird, CDA, RDH, MA

Participants:

Samantha McFeely
Keefe McFeely, Jr
James Barnthouse
Gar Chan, DDS

Contents

Preface

We invite the dental clinician to learn and review dental digital radiography technique. The potential rewards of digital radiography are great. Digital dental x-rays are chemical free, use a low dose of radiation, and have excellent image quality. Moreover, records are archived electronically, allowing for easy access and transfer. Digital techniques also allow use of the 'Logicon' system for caries detection enhancement.[1] This system, in conjunction with the sensitivity of digital sensors, allows a radiation dose reduction of thirty to fifty percent relative to the dosage received using standard radiological techniques.[2]

The clinician will adapt to the presence of computer infrastructure and software: a fiber optic cord, or cordless sensor, in the operatory; the parallel placement of the dental detector sensor; and the detailed use of DENTSPLY RINN style holding devices, in order to expeditiously produce images for diagnostic digital radiography in real time. As 14% or less of the dentists in the United States are using digital radiography as of the year 2005,[3] this booklet is designed to address the dental clinician or student who needs to learn or review general dental radiography. Topics covered will include computer management, radiography technique with digital sensor holders, and criteria for patient care and image evaluation.

Dental professionals can expect continuing change, and should continue to update dental radiology education.

---

[1] **Gakenheimer, DC.** The Efficacy of a Computerized Caries Detector in Intraoral Digital Radiography. IL: JADA, Vol 133, July 2002, 889.

[2] **Annals of the ICRP, International Commission on Radiological Protection: Managing Patient Dose In Digital Radiology.** Sweden: Elsevier. Pub 93 Vol. 34, No.1, 2004, 16.

[3] **GSC Home Study.** Dietz, E. 541: Dental Radiography and Digital Imaging. Sacramento: 2005, 3.

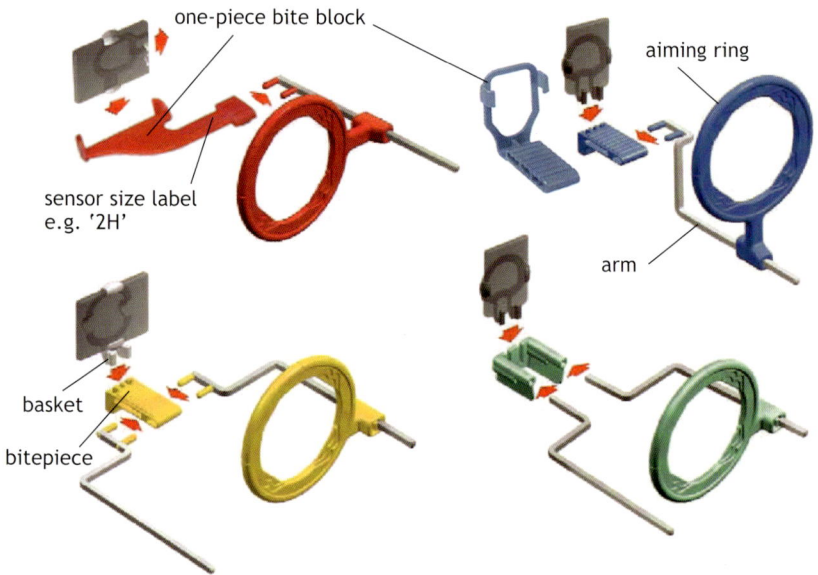

one-piece bite block

aiming ring

sensor size label
e.g. '2H'

arm

basket

bitepiece

Figure 1 *Assembling the color coded modular components XCP-DS ®*

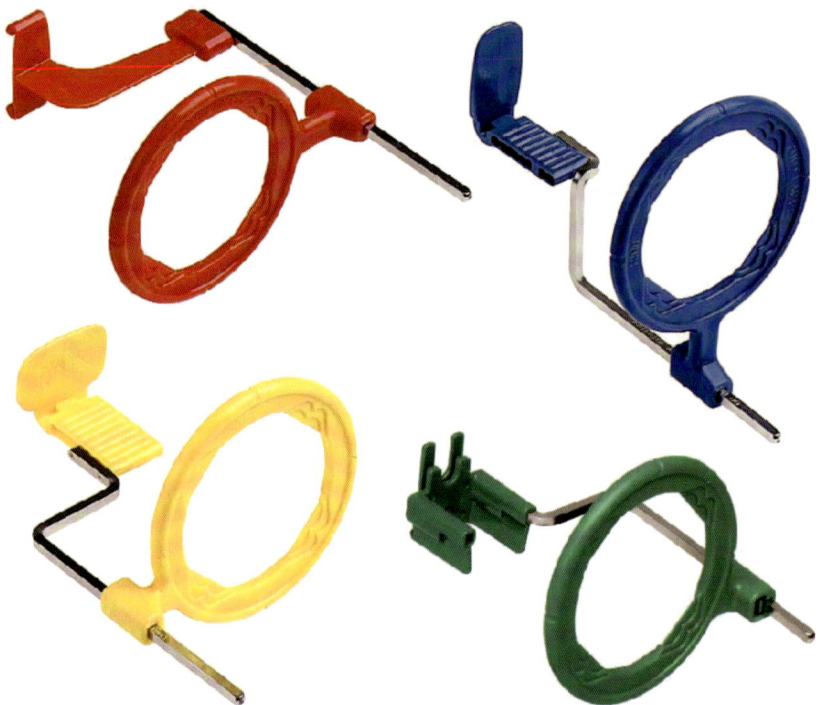

Figure 2 *The assembled instrument holders XCP ®*

Chapter 1

Armamentarium Guide for Radiography Components

Purpose

The purpose of this radiography publication is to detail the parallel placement of Digital Sensor (DS) holders. Like solving a math problem, we go back to the basics of one step at a time, and a picture for every step. Areas discussed will be the selection of RINN style holders, sensor effectiveness, patient care, optimum diagnostic quality, and basic radiation safety.

Angle

A dental Digital Sensor (DS) is designed to capture an image of two of the 206 bones in the body, the maxillary and the mandibular arch. The technique to ensure the proper angle at which to capture radiographic images is known by various names, such as Paralleling Technique, Long Cone Technique, Right Angle Technique, Extension Cone Paralleling (XCP®) and Extension Cone Paralleling–Digital Sensor (XCP–DS®). Long cone refers to increased Source to Image Distance or SID. RINN style holders are key to the *parallel* technique with digital sensor, in order to present objects in their correct size, location, and with accurate angulation.

Digital Radiography

Computer technology has led to an electronic sensor imaging system know as "Digital Radiography."[4] Digital radiography requires conventional equipment to capture diagnostic roentgen rays and uses computerized accessories to record the image and displays radiography information. Digital radiography captures the image, and passes the information to the computer. Digital images exist as bits of binary information stored and archived in a computer file until the image is placed on a viewing device such as a computer monitor or printed on paper. On a computer screen, the image is made up of small rectangular picture elements,

---

[4] **Torres & Ehrlich, Bird DL.**, Robinson DS. Modern Dental Assistant. 7th ed. St Louis, Missouri: Elsevier Sanders Inc, 2005, 677.

known as pixels. Each pixel represents one color or shade of gray. The computer can produce an image instantly, in real time.

The clinician will also need a computer station, and a network connecting the front and back office. Digital technology provides instant educational visuals and is already widely used in such places as hospitals, fax machines, and Digital Video Discs (DVD's).

DICOM (Digital Imaging and Communications in Medicine) has been developed as a common standard for digital radiography. Equipment and software which adhere to this standard allows the transfer of images between offices that utilize equipment from various manufacturers.

EQUIPMENT

The equipment utilized in this booklet is the DENTSPLY RINN style holders. The sensor specifications vary by planar dimensions, sensor type or size, pixel size, external dimensions, image active area, pixel area, number of pixels, thickness, storage and operating temperature range, capture card and computer specifications. See Table 1. The size 1 sensor is used for anterior teeth and pedodontic patients. The size 2 sensor is used for posterior teeth and adult patients.

SENSOR TYPE OR SIZE	PIXEL SIZE	EXTERNAL DIMENSIONS	IMAGE AC-TIVE AREA	PIXEL AREA	NUMBER OF PIXELS
1	22 x 22	39.5 x 26.0	30.1 x 20.1	1368 x 912	1,247,616
2	22 x 22	42.5 x 32.5	36.1 x 27.5	1640 x 1250	2,050,000

Table 1 *e2v technologies Sample Sensor Specifications. (Source: Yoon, DC., Chen, A. Users Manual for XDR Version 2.3 Cyber Medical Imaging (CMI), 2005).*

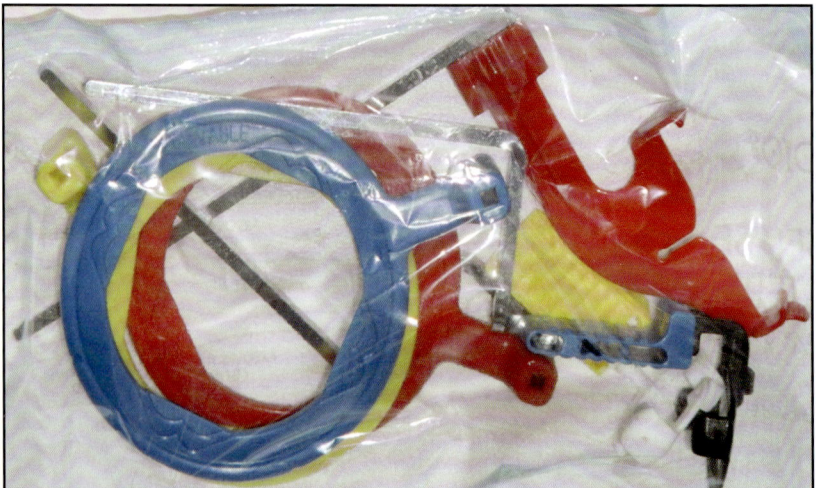

Figure 3 *Organize the operatory kit into cases*

ORGANIZE AND STERILIZE

Take the XCP-DS® kit out of the box and organize the pieces into sterilized surgical packages, as shown in Figure 3. Sterilize the pieces while they are in the surgical package, and they will then be ready to put onto the proper tray set-up.

INSTRUMENT ASSEMBLY

Figures 1 and 2 (see page 8) show the instrument holder parts, including aiming rings, indicator arms, color-coded bite blocks, bite-wing bite blocks, and bitepieces. Some brand specific sensors use a variety of shapes called names such as 'basket', or 'one-piece bite block'. Assemble the color coded parts in order to produce the proper instrument. With the one-piece bite block, the sensor passes through the bite block hole.[5] The sensor can be seen through the aiming ring when the instrument is correctly assembled. This allows for quick alignment when the holder is centered and aligned with the x-ray head, or Position Indicator Device (PID).

Follow the instrument part instructions and insert the two pronged indicator arm into the two far **end holes** of the bite block, away from the sensor slots.

---

a INSERT the standardized aiming ring into the stainless steel arm.

b SELECT the corresponding color coded one-piece bite block or bitepiece.

c ATTACH the very end of bite block or bitepiece to stainless steel arm pins.

d AFTER treatment, dismantle the instruments and autoclave the XCP-DS® instrument per manufacturer's instructions.

Table 2 *Summary of the instrument assembly*

DIGITAL SENSOR ASSEMBLY

Insert the sensor into a plastic sleeve barrier. A finger cot may be added over the disposable sheath for added stability and to ensure the prevention of cross-contamination.[6] The barrier protects the sensor from pathogenic microorganisms. Proper care of the sensor involves a disposable barrier cover, and following the manufacturer's precise directions for sensor disinfection. Avoid cord kinks and heat. Use extreme care when removing the plastic barrier from the sensor: slip the barrier off from the tip in order to avoid damaging the cord and bundle area. Place cord end of the sensor into an anti static bag when not in use. The barrier technique is also practiced for covering computer parts, including the keyboard and Universal Serial Bus (USB) port and plug.

---

[5] **XCP-DS**® Instructional Poster. Elgin, IL: DENTSPLY RINN, #5566, 2005.
[6] **Hocket SD, Honey JR, Ruiz F, Baisden MK, Hoen MM.** Assessing the Effectiveness of Direct Digital Radiography Barrier Sheaths and Finger Cots. JADA; V131, No. 4, 2000, 466.

THE PATIENT

Locate the patient file in the computer application. Before beginning the procedure ask the patient permission to take the necessary x-rays and ask about any dental concerns. Explain the purpose of the visit, and answer any questions about digital radiography. Drape the lead apron, and then drape the patient bib.

Show the patient the RINN holder set-up. Explain where they will bite down, and that they may feel the sensor press against the roof and the floor of their mouth. Ask the patient to swallow, in order to relax the facial muscles, and to breathe through their nose.

CONSISTENCY

The digital sensor (DS) fits into protective sleeves and uses a holding device to ensure proper placement. It is here where consistency counts. The clinician holds the DENTSPLY RINN instrument by the aiming ring, as shown in Figure 4. Face the flat side of the barrier covered sensor into the bite-piece, facing

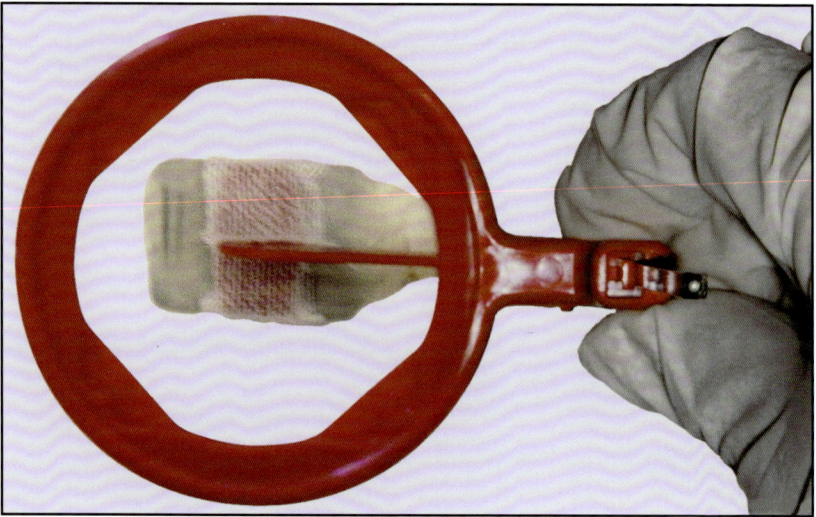

Figure 4 *Hold assembled instrument by the end of the aiming ring. Showing barrier sleeve, covered sensor with taped XCP® and finger cot cover. For XCP-DS® go to page 46.*

the sensor forward towards the white light x-ray beam and aiming ring. If the flat side of the sensor is facing the wrong direction, then the captured image will appear with wire lines at the cord end. The sensor cord end is outward, toward the midline, alongside the stainless steel arm.

Instruct the patient to position his or her neck in the head rest. Make sure the patient is comfortable. Ask the patient, for example, "How does the head rest feel?" To ease the weight of the cord, adjust the sensor cord to loop around the lead thyroid collar and bib.

Activate the x-ray icon in the computer application before sensor placement. The operator can use the same stance with every placement of the sensor and the long cone x-ray tube head (the PID or Position Indicator Device). For example, the clinician stands by the patient's knees and is consistent with this stance with every placement. There is no need to walk around to the other side of the chair: all you need to see is visible from this stance by the patient's knees. In order to rotate and move the PID, place one hand on the cone cylinder and the other hand on the PID back. After all of the requested images are captured, the clinician follows the manufacturer's instructions to 'save' the radiographs to the patient file. The dentist and patient are waiting for timely information; therefore, the dental clinician should be as efficient as possible.

Develop a set pattern of movements including hand and eye movements, set up, exiting the treatment room (remembering to avoid the cord cable line), and capture. Follow the same set routine, with all patients, under all circumstances. This consistency will enable you to efficiently produce quality radiography.

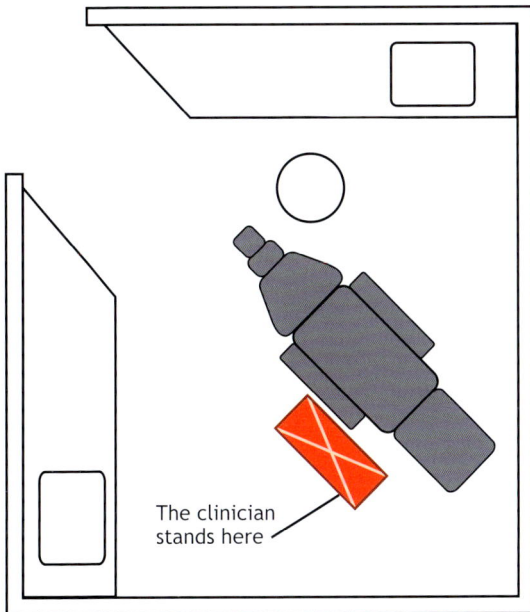

The clinician stands here

Figure 5 *Board Stance: "X Marks the Spot"*

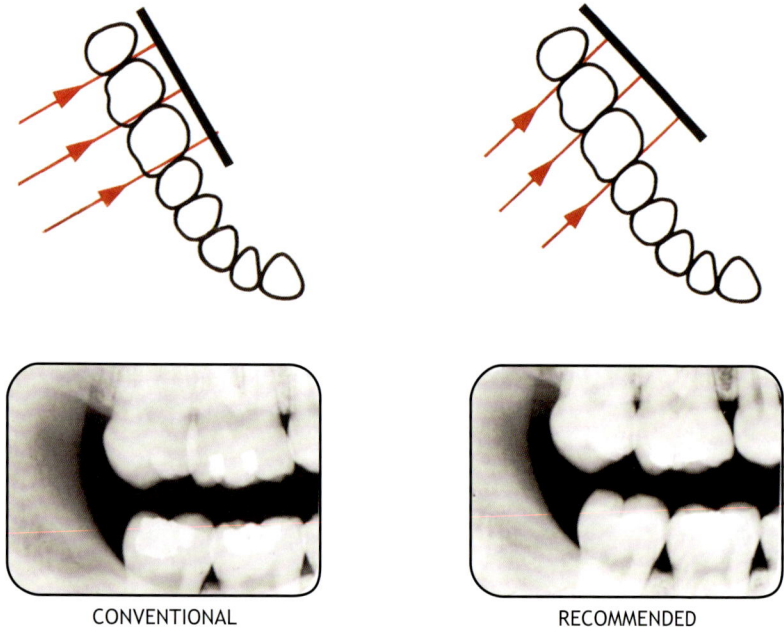

CONVENTIONAL

Since the interproximal surfaces of the molar teeth are in a mesio-distal relationship to the midsaggital plane, conventional positioning of the sensor parallel to the midsaggital plane or to the buccal surfaces results in overlapping of the contact areas and closure of the interproximal spaces.

RECOMMENDED

To avoid this condition, it is recommended that the sensor be positioned perpendicular to the interproximal spaces. This requires a diagonal placement of the sensor with the anterior border at a greater distance than the posterior border from the lingual surface of the teeth.

Figure 6ª *Recommended Bitewing Exposure Angle*
(Courtesy of DENTSPLY RINN: Intraoral Radiography with RINN XCP/ BAI Instruments. Elgin IL: RINN Corp., form # 1245-289, 1989, 36).

Chapter 2

How to Capture the Radiographic Image

Basic Steps

Steps **1** and **2** below have been previously discussed. Step **3** in particular will vary, depending on the type of radiograph:

1 **LOCATE** the patient file and open the x-ray application in the computer.

2 **ASSEMBLE** the RINN style instrument, insert barrier covered sensor on bite block.

3 **POSITION** instrument for BWX, anterior PA, posterior PA, Occlusal, or the RCT image.

4 **INSTRUCT** patient to gently close, then center and align x-ray tube PID on aiming ring.

5 **CAPTURE** the radiographic image by activating x-ray button, and **SAVE** to records.

Variations of these basic steps are outlined in this next section. Positioning in the mouth for the bitewing exposure is detailed, followed by the anterior periapical, posterior periapical, occlusal, and endodontic periapical.

BITEWING EXPOSURE (BWX)

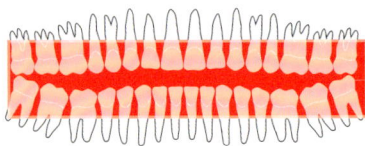

Color: *Red*
Numbers: *1-32*
Positions: *Horizontal BWX, Vertical BWX;*
Central Incisor, Lateral Incisor,
Cuspid, Bicuspid, Molar

The BWX examination may reveal interproximal caries. XCP® and XCP-DS® bitewing holders provide stability, a standardized technique, and establishes an accurate examination record. Select the appropriate horizontal or vertical bite block, as instructed by the dentist. For the horizontal bitewing, rest the BWX bite block on the occlusal surface of the mandibular teeth, as shown in Figures 6b and 6c. For the vertical bitewing, roll the sensor past the lower incisors in order to insert the sensor into the patient's mouth, placing the cord end toward the palate (see Figure 6d).

For added comfort, the RINN XCP-DS® bitewing holder may need, depending on the patient, the added use of products such as Edge-Ease™ to tape, capture, and stabilize the bitewing holder. Depending on the individual patient the clinician may modify the interproximal molar BWX technique by adding additional molar BWX, bicuspid BWX, cuspid BWX, or central/lateral BWX images.

Because the doctor will use the bitewing images to diagnose tooth decay, it is very important to ensure an open contact interproximal space, meaning that the images of each tooth do not overlap. Place the sensor perpendicular to the interproximal space such that the anterior distance is greater than the posterior border on lingual. The sensor with holder is simply placed parallel, right next to the molar or bicuspid on the lower arch between the tongue and the teeth, and then the anterior angle is opened slightly, as shown in Figure 6ª on page 14.

MOLAR	BITEWING

1 **LOCATE** the patient file and open the x-ray application in the computer.

2 **ASSEMBLE** bitewing instrument and insert barrier covered sensor on bite block. Face flat side of barrier covered sensor forward toward aiming ring, fit into bitepiece.

3 **POSITION** instrument for BWX with bite block resting on occlusal surfaces of mandibular teeth, align anterior boarder of the sensor with distal portion of second bicuspid. Open the anterior angle slightly (as shown in Figure 6ᵃ on page 14).

4 **INSTRUCT** patient to gently close, then center and align x-ray tube PID on aiming ring. Align PID unit in close proximity to the aiming ring.

5 **CAPTURE** the radiographic image by activating x-ray button, and **SAVE** to records.

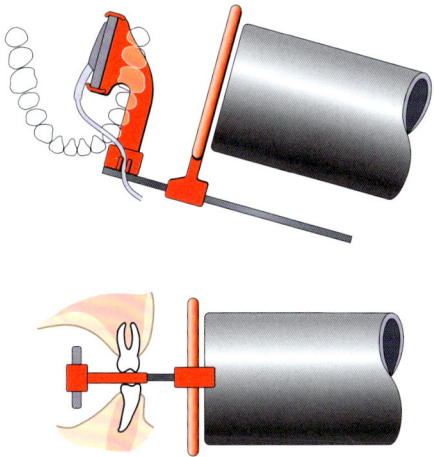

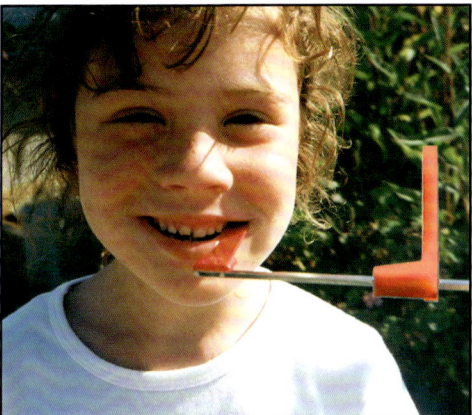

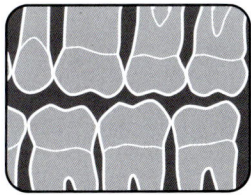

Figure 6b *Horizontal Molar Bitewing Exposure*

BWX

BICUSPID / CUSPID	BITEWING

1 **LOCATE** the patient file and open the x-ray application in the computer.

2 **ASSEMBLE** bitewing instrument and insert barrier covered sensor on bite block. Face flat side of barrier covered sensor forward toward aiming ring, fit into bitepiece.

3 **POSITION** instrument for BWX with bite block resting on occlusal surfaces of mandibular teeth, align anterior boarder of the sensor with distal portion of mandibular cuspid. Open the anterior angle. Modify for the cuspid image.

4 **INSTRUCT** patient to gently close, then center and align x-ray tube PID on aiming ring. Align PID unit in close proximity to the aiming ring.

5 **CAPTURE** the radiographic image by activating x-ray button, and **SAVE** to records.

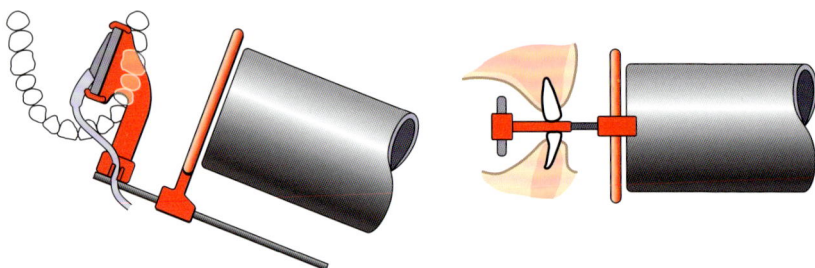

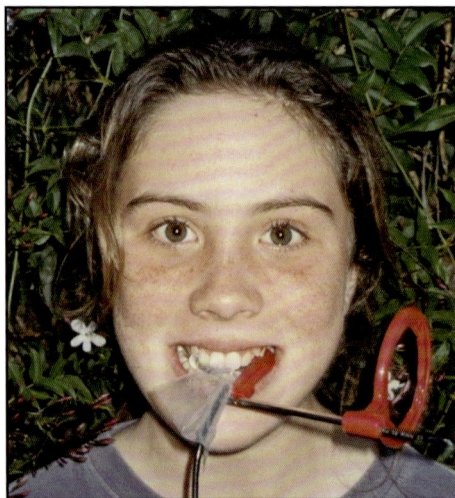

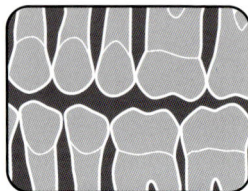

Figure 6c *Horizontal Bicuspid/Cuspid Bitewing Exposure*

CENTRAL / LATERAL INCISOR   **BITEWING**

BWX

1 **LOCATE** the patient file and open the x-ray application in the computer.

2 **ASSEMBLE** bitewing instrument and insert barrier covered sensor on bite block. Face flat side of barrier covered sensor forward toward aiming ring, fit into bitepiece.

3 **POSITION** instrument for BWX with bite block resting on the incisal surfaces of mandibular teeth, align center of sensor with the mesial of central or lateral incisors.

4 **INSTRUCT** patient to gently close, then center and align x-ray tube PID on aiming ring. Align PID unit in close proximity to the aiming ring.

5 **CAPTURE** the radiographic image by activating x-ray button, and **SAVE** to records.

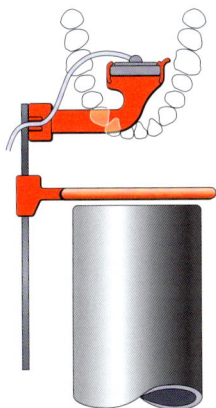

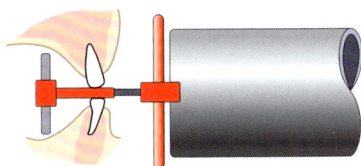

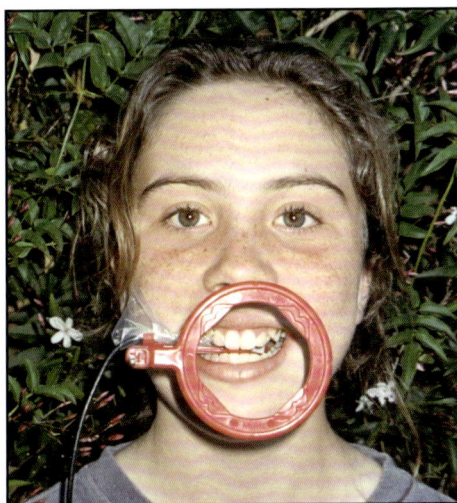

Figure 6d *Vertical Central/Lateral Incisor Bitewing Exposure*

ANTERIOR PERIAPICAL (PA)

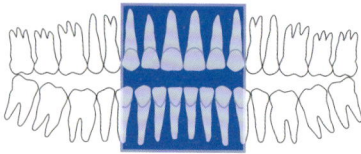

Color: *Blue*
Numbers: *6, 7, 8, 9, 10, 11*
27, 26, 25, 24, 23, 22
Positions: *Central Incisor, Lateral Incisor, Cuspid*

The patient is instructed to bite on the end line of the bite block or bite piece for anterior images of the centrals, laterals and cuspids on the mandibular and maxillary teeth.

The sensor image active area is smaller than the external dimension of the sensor, especially at the cord end bundle attachment area (see Table 1). For the anterior periapicals, as a result, the image may be obscured on the incisal edge. For the incisal edge of the anterior teeth, placing or taping a cotton roll on the bite block raises the sensor such that the incisal image is captured.

Instruct the patient to close near the end of the bite block rather than close to the sensor. The greater distance allows the entire tooth to be captured in one image. See Figure 7ᵃ.

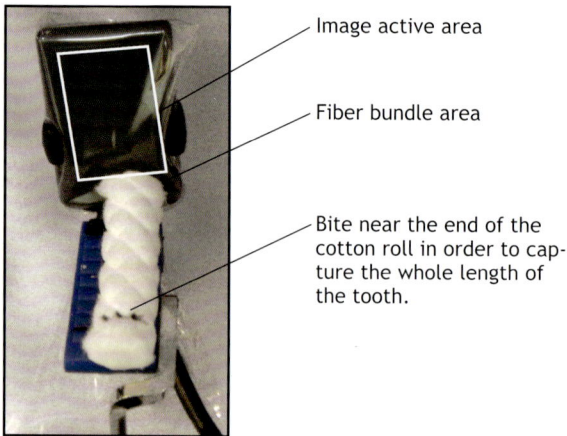

Image active area

Fiber bundle area

Bite near the end of the cotton roll in order to capture the whole length of the tooth.

Figure 7ᵃ *Fiber bundle at the cord end.*

MAXILLARY CENTRAL INCISOR	ANTERIOR PERIAPICAL

1 **LOCATE** the patient file and open the x-ray application in the computer.

2 **ASSEMBLE** anterior instrument, and insert the barrier covered sensor on bite block. Face flat side of barrier covered sensor forward toward aiming ring.

3 **POSITION** instrument for anterior PA with bite block and cotton roll resting on the incisal surfaces of maxillary teeth. Center sensor with the central incisors, distal palatal location, and parallel with the long axes of incisors. Entire horizontal length of the bite block is used to position sensor as posteriorly as possible.

4 **INSTRUCT** patient to gently close, then center and align x-ray tube PID on aiming ring.

5 **CAPTURE** the radiographic image by activating x-ray button, and **SAVE** to records.

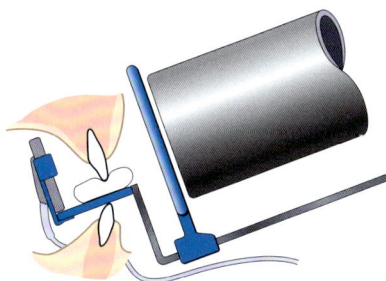

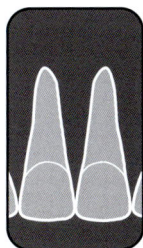

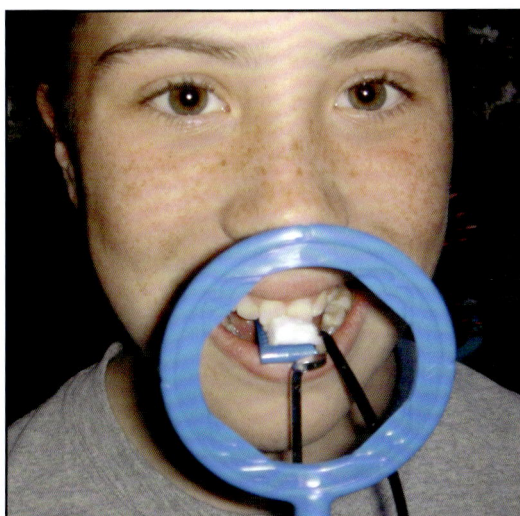

PA

Figure 7b *Maxillary Central Incisor Anterior Periapical*

MAXILLARY LATERAL INCISOR	ANTERIOR PERIAPICAL

1 **LOCATE** the patient file and open the x-ray application in the computer.

2 **ASSEMBLE** anterior instrument, and insert the barrier covered sensor on bite block. Face flat side of barrier covered sensor forward toward aiming ring.

3 **POSITION** instrument for anterior PA with bite block and cotton roll resting on the incisal surfaces of maxillary teeth, center sensor with the lateral incisors. Entire horizontal length of the bite block is used to position sensor as posteriorly as possible.

4 **INSTRUCT** patient to gently close, then center and align x-ray tube PID on aiming ring. Align PID unit in close proximity to the aiming ring.

5 **CAPTURE** the radiographic image by activating x-ray button, and **SAVE** to records.

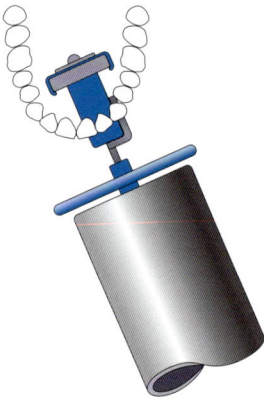

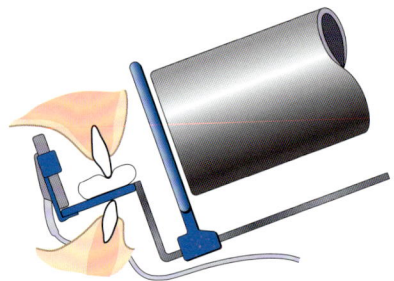

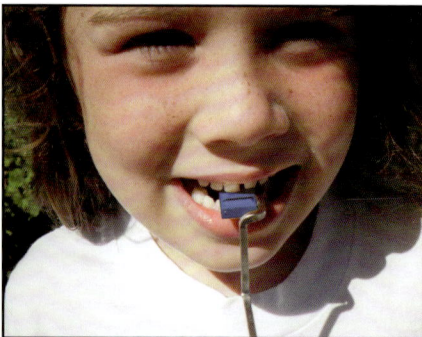

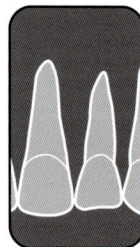

Figure 7c *Maxillary Lateral Incisor Periapical*

MAXILLARY CUSPID	ANTERIOR PERIAPICAL

1 **LOCATE** the patient file and open the x-ray application in the computer.

2 **ASSEMBLE** anterior instrument, and insert the barrier covered sensor on bite block. Face flat side of barrier covered sensor forward toward aiming ring.

3 **POSITION** instrument for anterior PA with bite block and cotton roll resting on the incisal surfaces of maxillary teeth, center sensor with the cuspid. Entire horizontal length of the bite block is utilized to position sensor as posterior as possible.

4 **INSTRUCT** patient to gently close, then center and align x-ray tube PID on aiming ring. Align PID unit in close proximity to the aiming ring.

5 **CAPTURE** the radiographic image by activating x-ray button, and **SAVE** to records.

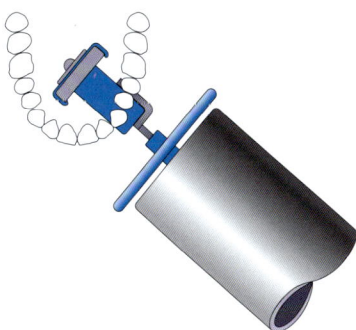

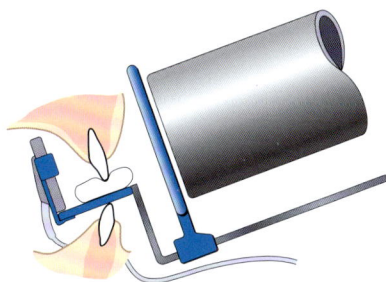

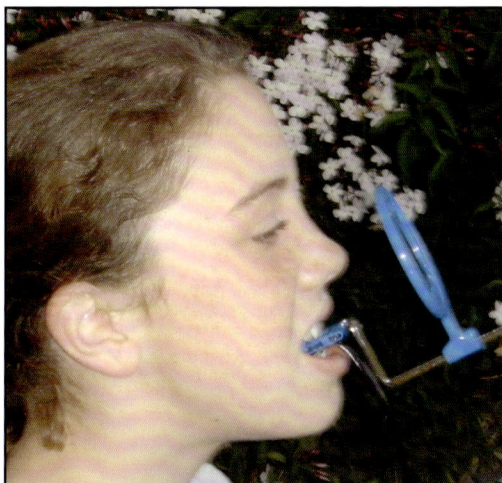

PA

Figure 7d *Maxillary Cuspid Periapical*

1 **LOCATE** the patient file and open the x-ray application in the computer.

2 **ASSEMBLE** anterior instrument, and insert the barrier covered sensor on bite block. Face flat side of barrier covered sensor forward toward aiming ring.

3 **POSITION** instrument for PA with bite block and cotton roll resting on the central and lateral incisal surface of mandibular teeth. Center sensor with the central/lateral incisors, parallel with long axes of incisor, on the lingual side. Entire horizontal length of the bite block is utilized to position as posterior as possible. The sensor may be placed on top of the tongue.

4 **INSTRUCT** patient to gently close, then center and align x-ray tube PID on aiming ring. Align PID unit in close proximity to the aiming ring.

5 **CAPTURE** the radiographic image by activating x-ray button, and **SAVE** to records.

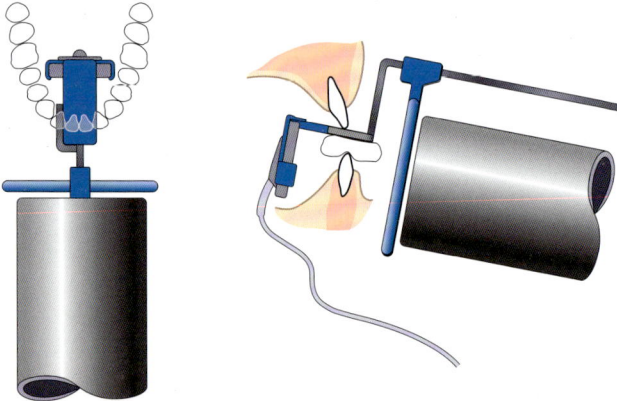

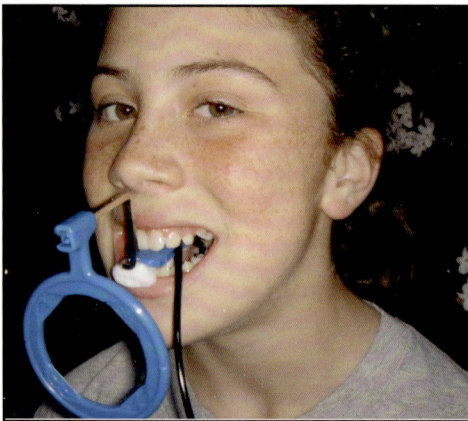

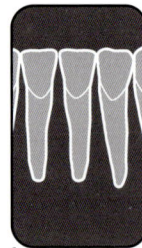

Figure 7e *Mandibular Central/Lateral Incisor Periapical*

MANDIBULAR CUSPID	ANTERIOR PERIAPICAL

1 **LOCATE** the patient file and open the x-ray application in the computer.

2 **ASSEMBLE** anterior instrument, and insert the barrier covered sensor on bite block. Face flat side of barrier covered sensor forward toward aiming ring.

3 **POSITION** instrument for anterior PA with bite block and cotton roll resting on the incisal surfaces of mandibular teeth, center sensor with the cuspid. Rest bite block on mandibular cuspid with inserted cotton roll and place parallel with long axis of mandibular teeth. The entire horizontal length of the bite block is utilized to position as posterior as possible. The covered sensor may be placed on top of the tongue.

4 **INSTRUCT** patient to gently close, then center and align x-ray tube PID on aiming ring. Align PID unit in close proximity to aiming ring.

5 **CAPTURE** the radiographic image by activating x-ray button, and **SAVE** to records.

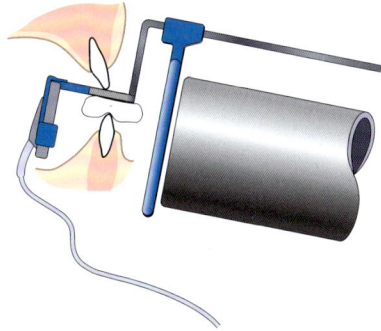

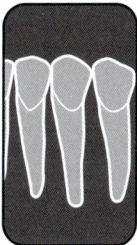

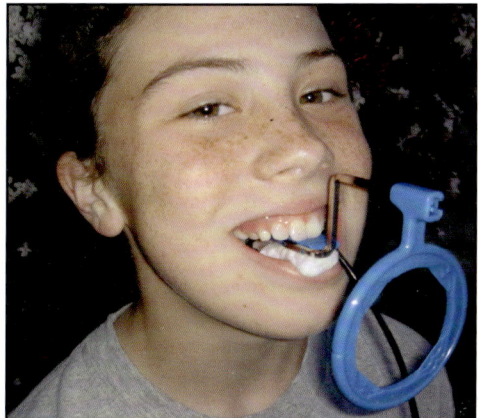

PA

Figure 7f *Mandibular Cuspid Periapical*

POSTERIOR PERIAPICAL (PA)

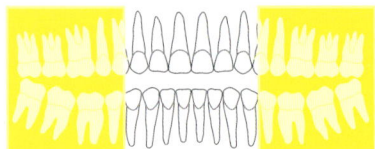

Color: *Yellow*
Numbers: *1, 2, 3, 4, 5; 12, 13, 14, 15, 16*
21, 20, 19, 18, 17;
32, 31, 30, 29, 28
Positions: *Molar, Bicuspid*

The posterior periapical set-up is more complex and requires some forethought to assemble. Attach the bite block to the arm, then attach the aiming ring such that you see the bite block when you look through the aiming ring. Next, attach the flat side of the sensor towards the aiming ring, and the cord end of the sensor towards the midline. The cord always runs alongside the arm.

For maxillary posterior teeth, the holder is placed on the **far** side of the patient's midline. As the patient bites down, the instrument rotates into parallel position. The increased distance enables the clinincian to capture the whole length of the tooth. In order to bring the RINN holder into the mouth, the cheek needs to be retracted. Start with the instrument angled upward toward the eye while placing the bite block on the occlusal teeth. Then rotate downward into parallel position. The clinician can then capture the images for one side of the upper teeth.

To save time, the same sensor assembly can then be used for the opposite lower side: simply rotate the entire assembly. For example, if the clinician starts with the upper left, the same assembly can then be used for the lower right.

On the lower arch, the RINN holder is placed in between the tongue and the teeth being captured. The cheek needs to be retracted (see figure 8e on page 28). In order to bring the holder into the mouth, angle the instrument down by the chin and rotate into position while placing the bite block on the occlusal surface.

The posterior arm is not symmetric; however, it is reversible. Rotating the assembly allows the clinician to capture the opposite quadrants. For instance, if the clinician started with the upper left and lower right quadrants, it is reassembled for the upper right and lower left. Disassemble the components, flip them over, and reassemble. Note that the yellow aiming ring is attached off center; therefore it also needs to be flipped and reassembled.

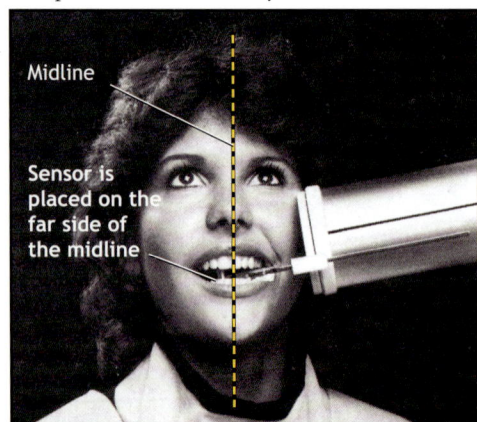

Figure 8ᵃ *Position the sensor to capture the entire length of the tooth. (Courtesy of DENTSPLY RINN)*

MAXILLARY MOLAR	POSTERIOR PERIAPICAL

1 **LOCATE** the patient file and open the x-ray application in the computer.

2 **ASSEMBLE** posterior instrument and insert the barrier covered sensor on bite block. Face the flat side of the barrier covered sensor facing forward toward aiming ring.

3 **POSITION** instrument for molar PA by centering sensor on second molar. Anterior edge of sensor is in vicinity of first molar and second bicuspid contact. Parallel sensor with long axes of molar which is achieved by lingual placement of sensor beyond midline of palate. Entire horizontal length of bite block is utilized to position sensor in mid-palatal area. Touch bite block on occlusal surface of maxillary molar. Cotton roll may be placed between the underside of bite block and opposing mandibular teeth.

4 **INSTRUCT** patient to gently close, then center and align x-ray tube PID on aiming ring. Align PID unit in close proximity to the aiming ring.

5 **CAPTURE** the radiographic image by activating x-ray button, and **SAVE** to records.

PA

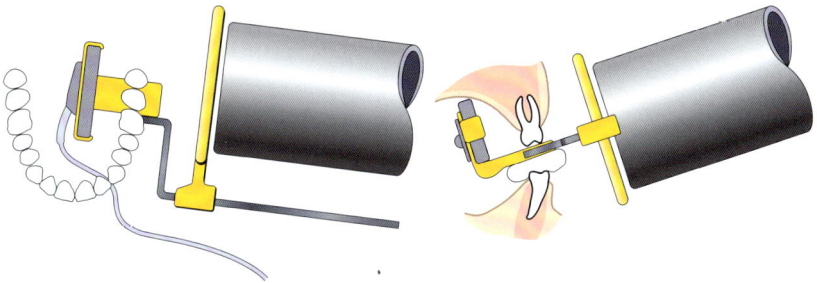

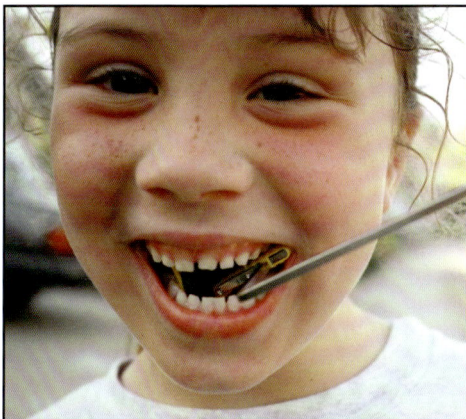

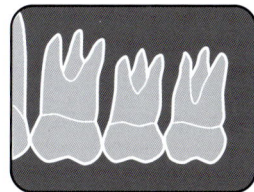

Figure 8b *Maxillary Molar Posterior Periapical*

1 **LOCATE** the patient file and open the x-ray application in the computer.

2 **ASSEMBLE** posterior instrument and insert the barrier covered sensor on bite block. Face the flat side of the barrier covered sensor facing forward toward aiming ring.

3 **POSITION** instrument for bicuspid PA by centering sensor on second bicuspid. Parallel sensor with long axes of bicuspid and place palatal to accomplish this relationship. Entire horizontal length of bite block is utilized to position sensor in mid-palatal area. The bite block is placed on the occlusal surface of maxillary teeth.

4 **INSTRUCT** patient to gently close, then center and align x-ray tube PID on aiming ring. Align PID unit in close proximity to the aiming ring.

5 **CAPTURE** the radiographic image by activating x-ray button, and **SAVE** to records.

PA

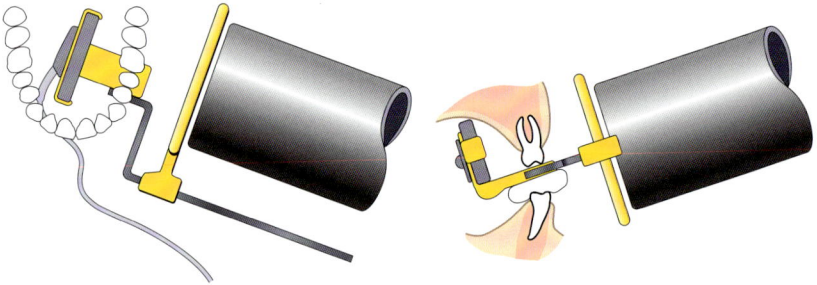

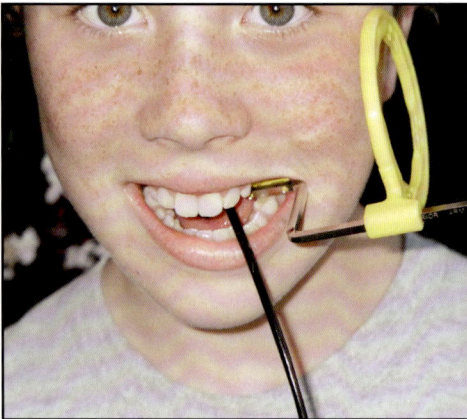

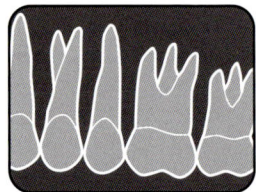

Figure 8c *Maxillary Bicuspid Posterior Periapical*

MANDIBULAR MOLAR	POSTERIOR PERIAPICAL

1 **LOCATE** the patient file and open the x-ray application in the computer.

2 **ASSEMBLE** posterior instrument and insert the barrier covered sensor on bite block. Face the flat side of the barrier covered sensor facing forward toward aiming ring.

3 **POSITION** instrument for molar PA by centering sensor on second molar. The anterior boarder of the sensor will align with the distal portion of the second bicuspid, first molar contact. Parallel sensor with long axes of molars. The bite block is rested on the occlusal surface of the mandibular teeth, right angles to long axes. The sensor is between tongue and teeth. Cotton roll may be placed on occlusal of maxillary.

4 **INSTRUCT** patient to gently close, then center and align x-ray tube PID on aiming ring. Align PID unit in close proximity to the aiming ring.

5 **CAPTURE** the radiographic image by activating x-ray button, and **SAVE** to records.

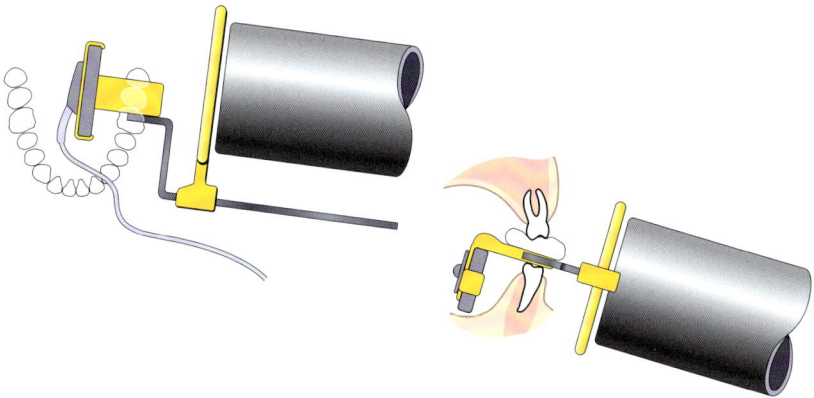

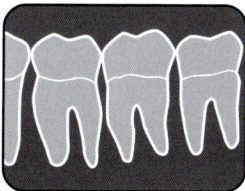

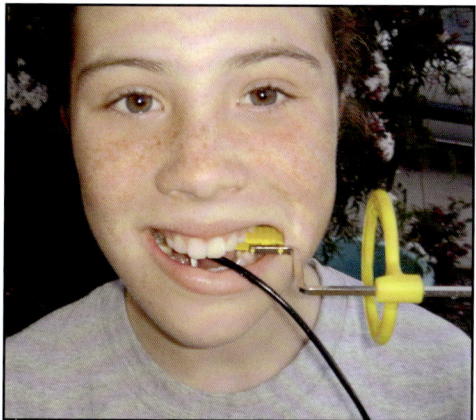

PA

Figure 8d *Mandibular Molar Posterior Periapical*

1 **LOCATE** the patient file and open the x-ray application in the computer.

2 **ASSEMBLE** posterior instrument and insert the barrier covered sensor on bite block. Face the flat side of the barrier covered sensor facing forward toward aiming ring.

3 **POSITION** instrument for bicuspid PA by centering sensor on bicuspid. The anterior boarder of the sensor will align with the distal portion of the cuspid. The bite block is placed on the occlusal surface of the mandibular teeth. The sensor is between tongue and teeth. A cotton roll may be placed on the maxillary occlusal surface.

4 **INSTRUCT** patient to gently close, then center and align x-ray tube PID on aiming ring. Align PID unit in close proximity to the aiming ring.

5 **CAPTURE** the radiographic image by activating x-ray button, and **SAVE** to records.

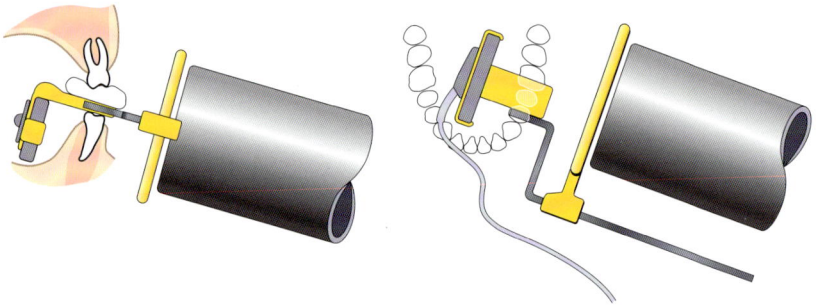

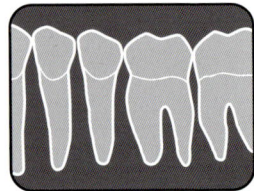

Figure 8e *Mandibular Bicuspid Posterior Periapical*

OCCLUSAL LOCATION (OCCL)

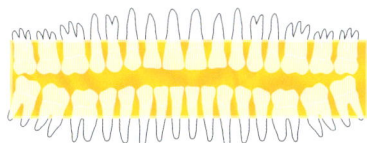

Color: *Yellow*
Numbers: *1-32*
Positions: *Maxillary Arch;*
Mandibular Arch

Occlusal views use the same yellow holder as the posterior periapical. The occlusal radiographs are used to obtain a wider area presentation to capture the position of all teeth in the arch. The occlusal image may reveal fractures or growths; help with diagnosis in trismus (locked jaw); and provide rapid surveys of teeth. The pedodontic occlusal radiograph uses less exposure time because the tissues are thinner. An occlusal (OCCL) image may be viewed in conjunction with bitewing exposures (BWX), and periapicals (PA) to localize buccolingual pathology.

To capture the image, position so that the central beam is directed perpendicular to the occlusal plane aligned with the center of the sensor and holder.

OCCL

MAXILLARY OR MANDIBULAR	OCCLUSAL

1 **LOCATE** the patient file and open the x-ray application in the computer.

2 **ASSEMBLE** posterior instrument, insert the barrier covered sensor on bite block. Face the flat side of the barrier covered sensor facing forward toward the aiming ring.

3 **POSITION** instrument for upper arch or lower arch image and direct the aiming ring to be perpendicular to the occlusal plane and aligned flat to the sensor.

4 **INSTRUCT** patient to gently close, then center and align x-ray tube PID on aiming ring. Align PID unit in close proximity to the aiming ring.

5 **CAPTURE** the radiographic image by activating x-ray button, and **SAVE** to records.

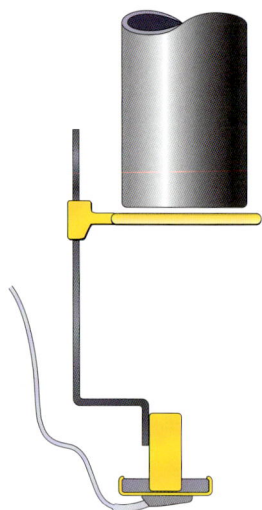

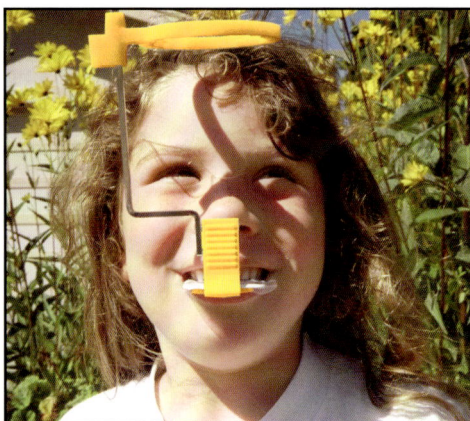

OCCL

Figure 9 *Occlusal view of maxillary arch*

ENDODONTIC PERIAPICAL (RCT)

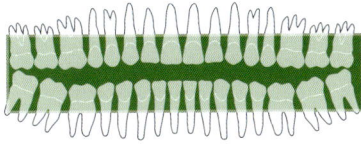

Color: *Green*
Numbers: *1-32*
Positions: *Central Incisor, Lateral Incisor, Cuspid, Bicuspid, Molar*

The green endodontic bite block is designed to rest on the adjacent teeth for those situations where the patient is unable to put pressure on the tooth being imaged. This bite block may be placed over a 'hot' highly sensitive tooth. It is also used during a root canal treatment.

As with the yellow (posterior periapical) holder, the posterior arm is not symmetric; however, it is reversible. Reversing the assembly allows the clinician to capture the opposite quadrants. For instance, if the clinician had the instrument assembled for use on the upper left quadrant, and needed radiographs of teeth on the upper right, the parts would need to be disassembled, flipped, and reassembled.

The root canal treatment (RCT) uses working images, captured at various stages of the treatment process. The working radiographs are used to determine the canal number, shape, location, size, and length.

Endodontic images are taken at two or three different angles to enable the dentist to view more than one canal in a multi-canaled tooth. To start, a mandibular molar RCT periapical, for example, may be taken straight-on showing an overlapping of canals.

For the next view, the holder will be angled slightly mesially (toward the patients midline). The mesial angled image will then separate the canals. In a two rooted bifurcated mandibular molar tooth, four canals may appear in the mesial view. At this angle, the mesial root will show the mesiolingual, followed by the mesiobuccal canal. The distal root will reveal the distolingual canal and the distobuccal canal. [7]

For the third view, the holder angle is opened distally. The clinician may capture a distal image by placing the sensor such that the anterior border is at a lesser distance, and the posterior border is at a greater distance, on the lingual. Then, close the anterior angle slightly by moving the aiming ring back by the patient's ear.

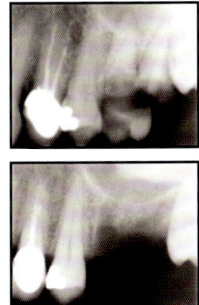

Figure 10ᵃ *Post operative root canal treatment. Endodontic images are taken at different angles in order to view multi-canaled teeth.*

[7] Finkbiner BL., Johnson CS. Comprehensive Dental Assisting. St Louis: Mosby, 1995, 823.

ROOT CANAL TREATMENT	ENDODONTIC PERIAPICAL

1 **LOCATE** the patient file and open the x-ray application in the computer.

2 **ASSEMBLE** endodontic instrument, insert the barrier covered sensor on bite block. Face the flat side of the barrier covered sensor facing forward toward the aiming ring.

3 **POSITION** instrument assembly for RCT over the tooth files and dental dam clamp, and rest the bite piece on the adjacent teeth with the dental dam frame released. Place the sensor so that it is adjacent to the tooth being captured and behind the dental dam. The Endo instrument can be used on a hot tooth and in all areas of the mouth.

4 **INSTRUCT** patient to gently close, then center and align x-ray tube PID on aiming ring. Align PID unit in close proximity to the aiming ring.

5 **CAPTURE** the radiographic image by activating the x-ray button, and **SAVE** to records.

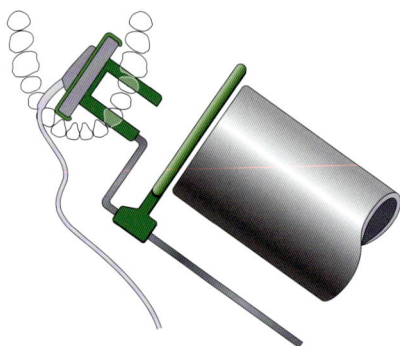

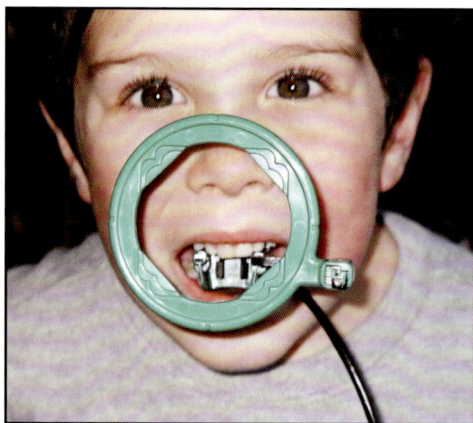

Figure 10b *Endodontic Periapical*

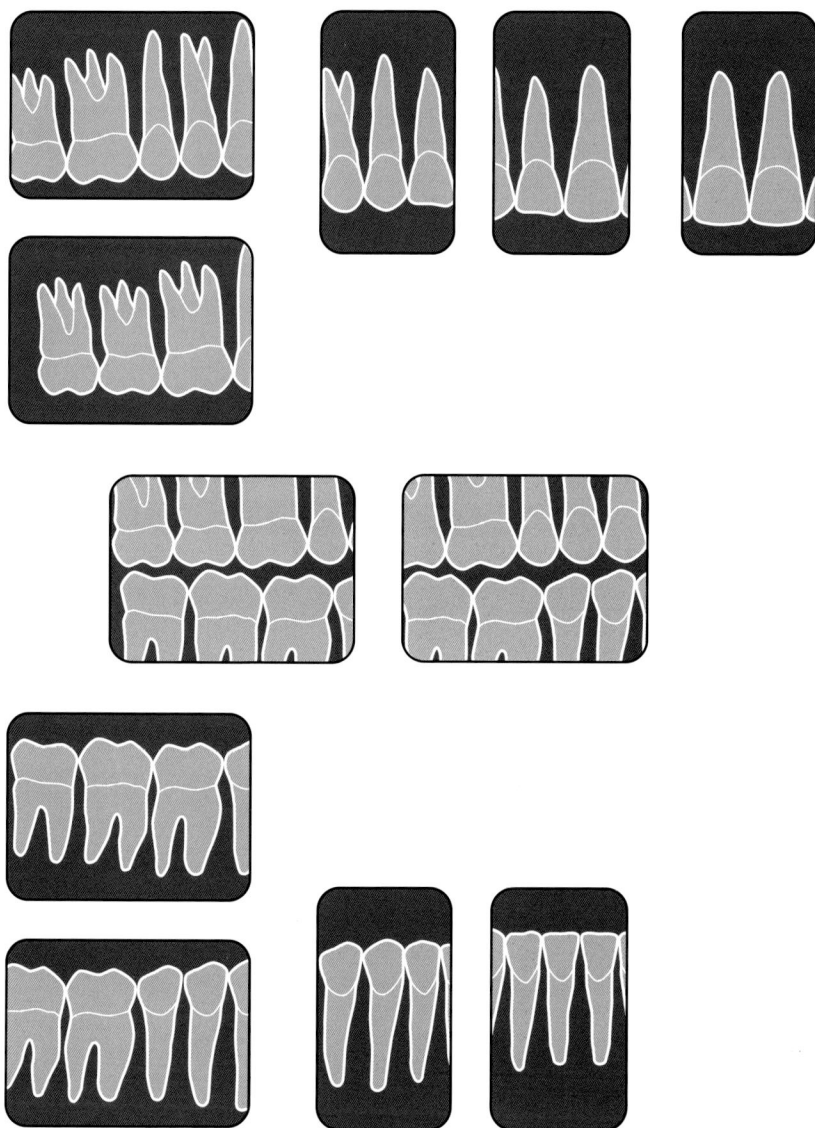

Figure 11 *Ideal Full Mouth Exposure (FMX) Series showing Class I occlusion (CL I)*

Figure 12 *Sample FMX format starting with BWX #1. Note that cord end is outward.*

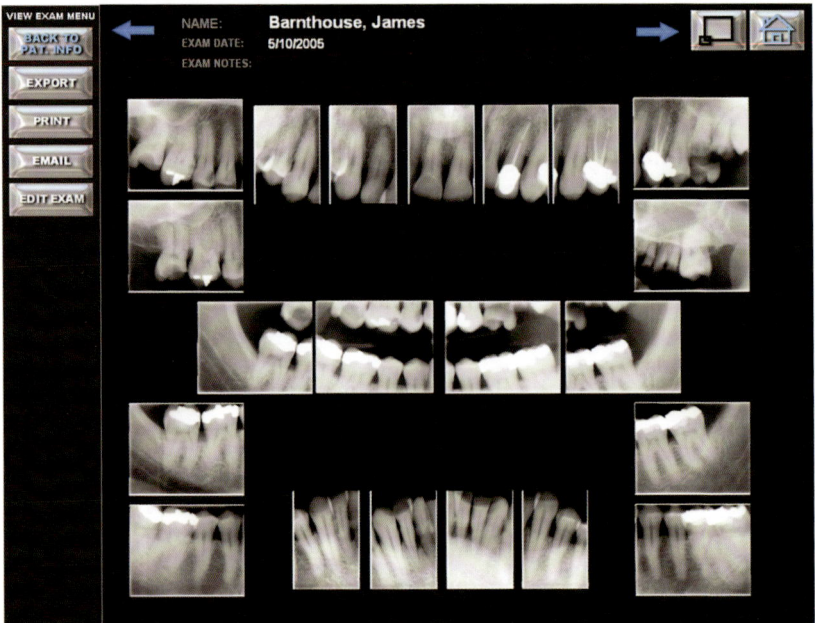

Figure 13 *FMX Series. Overview of the case:*
Class II / Division 2 molars and cuspids (right) – Complete anterior overbite – Mandibular and upper anteriors appear retro. – Lower incisors #23-26 = perio compromised

Chapter 3

CRITERIA FOR EVALUTATION AND ADDITIONAL CIRCUMSTANCES

CRITERIA FOR EVALUATION

The ideal Full Mouth Exposure (FMX) series views the alveolar crest, centers the occlusal plane, opens contacts for the bitewing, and captures the occlusal border 3 millimeters beyond the apices for periapicals. For excellent clinical evaluation, follow the ideal Full Mouth X-ray, as shown in Figure 11. In an ideal radiograph, the crown, root and apex are fully depicted with interproximal alveolar crests, contact areas, and surrounding osseous regions. The sensor is place parallel to the longitudinal axis of the tooth, and the central ray perpendicular to the tooth and sensor. The image of all teeth and other structures are shown in proper placement, relative size, and contour with minimal distortion and without overlapping images where anatomically possible.

IMAGE FORMATION

Density is the degree of darkness on a radiographic image. Decreasing the exposure time lowers the amount of radiation and lowers the density. *Contrast* is the difference in densities between adjacent areas of the radiographic image. [8]

ANGULATION

When an image is captured, the tooth, being a denser object than the surrounding tissue, absorbs some of the radiation, as if casting a shadow on the digital sensor. Like shadows cast by the sun, the angle of the x-ray beam, relative to the teeth and sensor, determines the vertical length of the radiographic image. The vertical angle determines how accurately the length of an object is reproduced. When the central ray is not perpendicular to the **sensor**, an *elongated* image will result. The tooth image will appear too long, like a late afternoon shadow. On

---

[8] Intraoral Radiography with RINN XCP®/BAI Instruments. Elgin IL: RINN Corp., form # 1245-289, 1989, 46-7.

the other hand, too much vertical angulation will result in a *foreshortened* image that will appear too short, like a midday shadow. In this case, the sensor is not placed parallel to the **tooth**.

A *cone cut* occurs when the central ray fails to be centered through the aiming ring, and the image is cut off by the edge of the x-ray tube PID, or cone. An *overlapping* image occurs when the central ray fails to be horizontally perpendicular to the sensor. In this case, adjacent teeth overlap in the captured image.

MODIFICATIONS
In order to adequately cover patients with long roots, place the sensor vertically rather than horizontally in the posterior instrument. A vertical bite block may be used. For molars, the clinician may angle the sensor slightly up, like an occlusal image.

For the edentulous patient, a cotton roll is substituted for the space normally occupied by the crown of the tooth. The holding instrument is positioned in the mouth with the sensor parallel to the ridge area being examined. The cotton roll or rolls is/are placed on the edentulous area. Position the sensor and use standard procedures to capture the image.

For the wisdom tooth, turn the aiming ring to open the distal space. The anterior border is at a lesser distance, and the posterior border is at a greater distance, on the lingual. Bring the aiming ring and PID back to where the ear meets the temporomandibular joint (TMJ), then aim the PID toward the midline.

ADULT AND PEDODONTIC PATIENT
The success of dentistry provided to the compromised patient depends on accuracy of sensor placement, and how well the dental team relates to the person in a positive, friendly, and honest manner. Explanations are integral parts of patient behavioral cooperation, education and modification. The patient can sense the clinicians' enjoyment of their work by watching facial expressions and listening to the tone of voice. Children need short timely appointments and enthusiastic attention.

RADIOBIOLOGY
What do we mean when we say that radiation is cumulative? The amount of radiation received today is added on to the amount received in all past exposures. The clinician must protect themselves from radiation exposure by standing behind a lead lined wall. Do not hold the sensor or PID during the exposure.[9] Radiation protection in the dental practice is achieved through utilization of a holding device which carefully aligns, positions, and stabilizes the sensor with the area to be captured.

---

[9] **Radiation Protection in Dental Practice.** Sacto,CA: Dept Health; Jan. 1987, 6.

Patient exposure is also reduced by the 'lead' apron. (The apron may be made of other materials as well). The clinician holds the apron by the shoulder flaps, and drapes the apron over the patient's shoulders. The sensor cord is draped on top of the bib, around the apron. Using a quality digital sensor system reduces radiation exposure. There is a professional responsibility and obligation to eliminate unnecessary radiation.

The standard for protection against radiation requires every exposure and release of radioactive material to be As Low As Reasonably Achievable (ALARA) and takes into account the technological improvements in radiation safety.

RULES AND REGULATIONS

The dentist and dental auxiliary are trained and state licensed for the use of x-rays. The clinician has passed an approved course which includes lecture, didactic laboratory, and clinical examination. The auxiliary becomes a licensed person who functions under the direct supervision of the licensed dentist. [10]

Currently there are proposed regulatory changes in California.[11] (In other states there may be similar bills, with similar intent). The proposed changes would require successful completion of a minimum of four full mouth surveys, of which no more than three sets may be completed using computer digital radiographic equipment, in order to receive an x-ray license.

DOCUMENTATION

The dental x-ray technician updates patients' health history, measures blood pressure, and documents routine questions and answers.[12] Typical questions might be "Does everything feel OK in your mouth?" or "Did you eat today?" [13]

A sample clinical notes format is as follows: Today's date, followed by the terminology "updated" or "no change," and then the type of appointment procedure, and the order performed. End with the signatures of the licensed individuals performing the treatment. For patients who require premedication, the dentist may also want the patient to sign the chart stating, for example "Took Premed X" followed by the signature.

---

[10] **Rossi, CH.**, Torres, HO. History Extended Functions Dental Auxiliaries. Chicago: ADAA, Vol. 68, No.3. 1999, 38.

[11] **Proposed Regulatory Language Barclay Official CA Code of Regulations Title 16 Division 10 #1014.** 1SF: Mar. 2005.

[12] **Martin, CH.**, Moore D.,Templin K., Moore J. Electronic Blood Pressure Monitor. Chicago: ADAH, 1998, 33.

[13] **Malamed, SF.** Pharmacology. CDA 2002 Fall Scientific Session. San Francisco, 2002, slide #27.

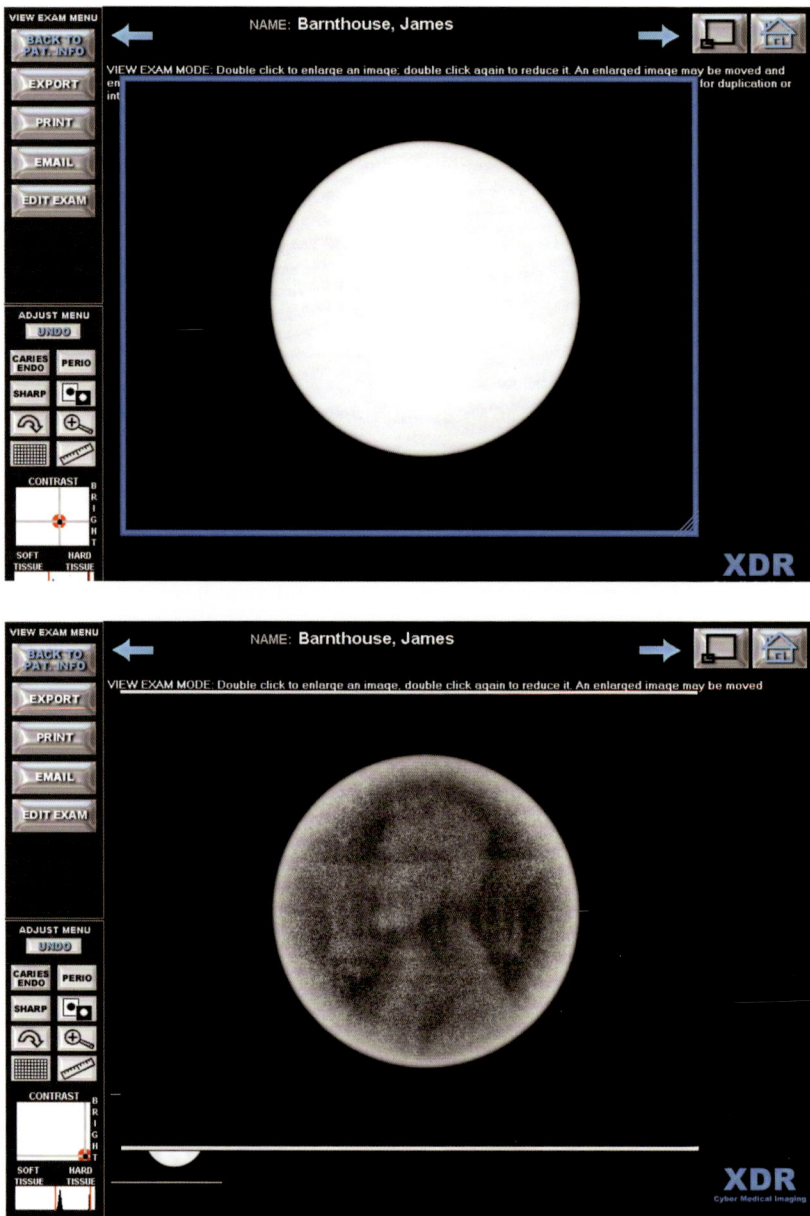

Figure 14 *The Coin Test. The XDR adjustment features are on the lower left corner. In the top image, the contrast controls have not yet been adjusted, and the features of the penny are not visible. Note that in the bottom image, the contrast slider has been moved to the lower right corner, bringing the data into a visible range.*

Chapter 4

Software for
Digital Radiography

Software

This section gives the clinician a brief introduction to software used with digital radiography. To quote Douglas C. Yoon, DDS, developer of the XDR digital imaging software, and the inventor of Logicon, an advanced caries detection system:

"XDR is easy to learn. The screen is designed to resemble a radiographic view box where radiographs are placed and manipulated by clicking and dragging the mouse. A radiograph can be taken with a single mouse click. Clear instructions on screen assist the operator at every step. Much like other Windows graphic interfaces, program control occurs through a two-step process with first selecting an object by moving the mouse and second, initiating events by clicking the left mouse button when the screen cursor (a small arrow) is over the appropriate screen item."

"Most of the tasks performed by XDR are initiated by a single left click. XDR provides ultra clear, high contrast film-like images while specialized soft tissue and hard tissue enhancement functions give the clinician the superior imagery needed for effective dental diagnosis. Clinicians should apply the standard of practice with respect to correct positioning geometry and exposure parameters and note the clear presence of standard anatomic landmarks."

"Database Link: XDR supports one way bridging from most practice management systems. This is an advanced feature. The purpose of bridging is to synchronize patient names in practice management software with patient names in XDR. Under this option the bridge is implemented with Data grabber utility supplied on CD-ROM." [14]

---

[14] **Yoon, DC.**, Chen, A. Users Manual for XDR Version 2.3 Cyber Medical Imaging (CMI), 2005, 39.

Quality digital systems set the standard for digital radiography to provide powerful diagnostic tools and advanced analysis algorithms. Image analysis includes enhanced caries and endodontal detection, periodontal enhancement, sharpness, inverse or reverse gray scale, rotation of an image, magnification, grid work, ruler measurement, contrast and brightness control, intensity range controls, and labeling and drawing features. Figures 12, 13, and 14 show typical XDR user interface screens.

QUESTIONS & ANSWERS WITH PROFESSOR YOON

What is the meaning of 'Logicon'?

'Logicon' was the name of the aerospace firm that I worked for when I developed the automated caries detector. Hence the product was named after the company.

What is D.C.A?

The distortion correction algorithm (DCA) is an algorithm that corrects the image for elongation only. That is, for distortion caused when the sensor is not placed perpendicular to the x-ray beam. It does not correct for foreshortening. That is, distortion caused when the tooth is not placed perpendicular to the x-ray beam.

What do we mean by gray scale?

'Gray scale' refers to the brightness range of a black and white digital image, such as digital x-rays.

How many shades of gray are there?

I'm sure you mean how many shades of gray are there in a digital image, because in the real world there is an infinite range of brightness intensities. Digital x-ray images are electronically captured with up to 4096 (12-bit) or 65,536 (16-bit) different shades of gray or brightness levels. The more levels you capture, the subtler the differences in brightness one can potentially detect. Electronic noise [15] and limitations in the detector scintillator screen and CCD technology create a practical limit of only about 100 to 200 detectible shades of gray. Thus all sensor systems are over sampling the data (which is OK; it's always better to over sample than under sample the data). To make matters even more complicated, computer monitors are only able to display 256 discrete shades of gray. Thus what part of the brightness range of the image data the program chooses to present onto the monitor becomes very important and is one of the important

---

[15] Electronic noise is analogous to auditory noise; in other words, minor background variations that result in visual noise in the x-ray image, giving the impression of graininess. (Source: Annals of the ICRP, International Commission on Radiological Protection: Managing Patient Dose In Digital Radiology. Sweden: Elsevier. Pub 93 Vol. 34, No.1, 2004, 9).

factors that determines the quality of all the different x-ray imaging systems. Sorry the answer is so complicated, but there is a lot of engineering going on here.

What is the highest resolution image possible?

Film is the sharpest medium and it can record 20-25 line pairs per millimeter (lp/mm).

How does the coin test show us an example of how the enhancement tools work?

The image of the two sides of the coin are compressed into a very narrow brightness range. Thus, it tends to appear as either pure white or black. I take the data in this narrow brightness range and do an extreem contrast stretch, thus revealing the image within this narrow range. It's sort of like playing with the brightness and contrast settings on your TV to better see detail in a washed out region of the image.

THE COIN TEST: REPORT ON ABRAHAM LINCOLN

Take an x-ray of an old 1960 or 1970 penny. Assemble the sensor in the yellow holder. Stand the x-ray holder up on a table with the flat side of the sensor facing upward. Place the penny with Lincoln's head facing downward, on top of the sensor. Align the PID to the aiming ring. Expose the image. Next, view the advanced program features and adjust the contrast, as pictured. The human eye can detect 32 distinct shades of gray.[16] Play with the controls of the Adjust Menu until you can see the Lincoln head. See how the contrast/brightness controls can bring the images into a visible range (refer to Figure 20 on page 40).

CLOSING

In closing, we give you this collection of digital radiography techniques. The dental practice conversion to digital radiography realizes the full potential and value of current technology. The practice of sensor placement technique and image aquisition will allow the clinical team to gain experience, save time, acquire confidence, and contribute to a smooth transition to the world of digital radiography.

---

[16] **Torres & Ehrlich, Bird DL.**, Robinson DS. Modern Dental Assistant. 7th ed. St Louis, Missouri: Elsevier Sanders Inc, 2005, 678.

REFERENCES

Anderson PC, Burkard MR. The Dental Assistant. 6th ed. Albany: Del Mar Pub, 1995.

Anderson PC, Clifford SB. Dental Radiology. Albany: Del Mar Pub Inc, 1982, 2-89.

Annals of the ICRP, International Commission on Radiological Protection: Managing Patient Dose In Digital Radiology. Sweden: Elsevier. Pub 93 Vol. 34, No.1, 2004.

Chasteen JE. Essentials of Clinical Dental Asstistant. 4th ed. St Louis: Mosby, 1989.

Caring For your Teeth: What Does Digital Radiography Mean? N.C: Sirona, 2005.

Condensed Version Intraoral Radiography With RINN XCP®. IL, RINN Corp, #1245c-0396.

Dalin, JB. The Bottom Line Digital Diag. Dental Economics: Penn Well Corp, Sept. 2004.

Digital Radiographs State-of-Art. Provo: Clinical Reasarch Asso Newsletter, Vol. 23, Sept. 1999.

Digital Radiography-2005. CRA Utah: CRA Foundation, Vol. 29, Issue 2, Feb. 2005.

Digital Radiography Congress Panel Discussion. The Image of Dentistry. Weisman, G., Christensen, G., Emmott, L., Flucke, J., Schiff, T. SF, CA: An Advanstar Exposition Production, April 30, 2005.

Digital X-ray: Sidexis/Networks Pocket Guide. NC: Sirona Dental Co, 2002.

Ehrlich A, Torres HO., Bird D. Essentials of Dental Assist. 2nd ed. Phil: Saunders, 1996.

Farman, H. A. Comparison of 18 Different X-ray Detectors currently used in Dentistry. Oral Surg, Oral Med, Oral Pathol, Oral Radiol, Endo D. MO: Elsevier, 2005; 99: 485-9.

Finkbiner BL., Johnson CS. Comprehensive Dental Assisting. St Louis: Mosby, 1995.

GSC Hm. Study. 541: Dental Radiography & Digital Imaging. Sacto. CA: 2005.

Gakenheimer, DC. The Efficacy of a Computerized Caries Detector in Intraoral Digital Radiography. IL: JADA, Vol. 133, July 2002, 883-889.

Hocket SD, Honey JR, Ruiz F, Baisden MK, Hoen MM. Assessing the Effectiveness of Direct Digital Radiography Barrier Sheaths and Finger Cots. JADA, Vol. 131, No. 4, 2000, 463-7.

Intraoral Radiography With RINN XCP®/BAI Instr. Elgin IL: RINN Corp., #1245-289, 1989.

J & J Secure-Comfort Cloth Tape, First Aid. Skillman, NJ: Johnson & Johnson, 2003.

Kehoe, B. Get Ready To Go Digital. Hygiene Report 2004 Products Guide. MN: Advan Star Communications, Oct. 2004.

Kehoe, B. Finally Filmless. Technology Guide 2005. Dental Practice Report. MN: Advan Star Communications, Vol 12, No. 10. Dec. 2004.

Kupfer P. Designing Products Based on Real Life / High-tech firms seek clues in anthropology. SF, CA: SF Chronicle, Jan 31, 2000, B.

Malamed, SF. Pharmacology. CDA 2002 Fall Scientific Session. SF, 2002.

Manley, V. Wrap-Ease, Edge-Ease. Corona, CA: Strong Dental Products, 619, 2005.

Martin, CH., Moore D., Templin K., Moore J. Electronic Blood Pressure Monitor. Chicago: ADAH, 1998, 33.

Miles, DA. Going Digital. American Fork, UT: Dentrix Dental Systems, Feb. 2004.

Olson, SS. Dental Radiography Laboratory Manual. Phil, PA: Saunders, 1995.

On Your Mind: Digital Dental X-Rays. NY: Consumer Reports On Health, July 2004: 12.

Proposed Regulatory Language Barclay Official CA Code Title 16, Professional and Vocational Regulations. Division 10. #1014. SF: Mar. 2005.

Radiation Protection In Dental Practice. State of California, Sacto, CA: Dept Health Services, Radiologic Health Branch, Jan. 1987.

Radiographic Technique & Safety, Taking Radiographs. Garden Grove: Medcom, 1992.

Rossi, CH.,Torres, HO. History Extended Functions Dental Auxiliaries. Chicago: ADAA, Vol. 68, No 3. 1999.

Schick No. 3; Achieve Optimal Result With Digital Radiography. JPH. NY: CDR Positioning System User Guide #B1051044/B1073080. Long Island City, NY: Schick Tech., 2003.

Sometimes Patients Are All Too Human: Scope Professional Service. S.N. 189-8842.

Tame your Sensor. Elgin, IL: DENTSPLY RINN, #MK90A, 2005.

Torres & Ehrlich, Bird DL., Robinson DS. Modern Dental Assistant. 7th ed. St Louis, Missouri: Elsevier Sanders Inc, 2002.

Torres & Ehrlich, Bird DL., Robinson DS. Modern Dental Assistant. 8th ed. St Louis, Missouri: Elsevier Sanders Inc, 2005.

West's Annotated California Code: Official CA Business and Professions Code Classification. Healing Arts Div. 2. St. Paul, MN: West Publishing Company. 1994.

Wheeler, RC. An Atlas of Tooth Form. Philadelphia: Saunders, 1969, 131-3.

White, SC., Yoon, DC., Tetradis, S. Digital Radiography in Dentistry: What It Should Do For You. CDA Journal, Vol. 27, No. 12. Dec. 1999.

Williamson, Gail F. Digital Radiography in Dentistry: Moving From Film Based to Digital Imaging 0409. ADAA, 2004.

XCP® **Instructional Poster.** Elgin, IL: DENTSPLY RINN, #5553, 2000, www.rinncorp.com

XCP-DS® **Instructional Poster.** Elgin, IL: DENTSPLY RINN, #55-1001, 2005.

XCP-DS® **Instructional Poster.** Elgin, IL: DENTSPLY RINN, #5566, 2005.

XCP-DS® **Product Catalog 2005.** Elgin, IL: DENTSPLY RINN, # MK01E, 2004; 15-19,32.

XCP-DS® **Guide to Perfect X-rays.** Elgin, IL: DENTSPLY RINN, #MK82, 2005.

XCP® **Radiography using the RINN XCP**®**/BAI Instruments.** Chesterland,OH: Academy of Dental Therapeutics and Stomatology, Director Florman,M., Lavi,E. #AS 8126; 2004.

XCP-DS® **Digital Sensor Positioning System and the RINN XCP**® **Endodontic Instrument Radiography using the RINN XCP**®**/BAI Instruments.** Chesterland, OH: Academy of Dental Therapeutics and Stomatology, Director Florman, M., Lavi, E. #RINN 1103.

X-Rays in Dentistry. Rochester: Kodak Publication Health Sci Mkt Div, No. D1-55,1985.

Yoon, DC., Chen, A. Users Manual For XDR Version 2.3 Cyber Medical Imaging (CMI), 2005.

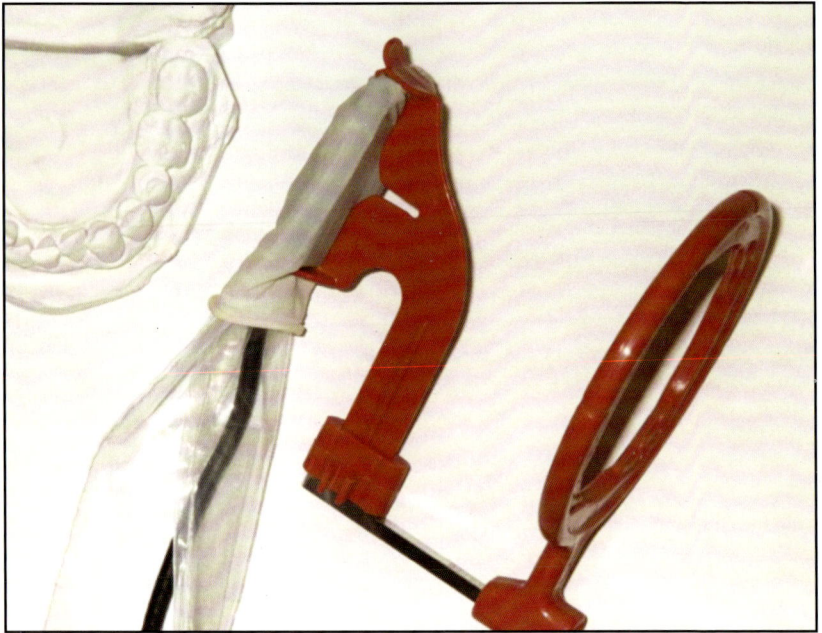